Water
219

河流新生机
Changing Rivers

Gunter Pauli

[比] 冈特·鲍利 著

[哥伦] 凯瑟琳娜·巴赫 绘

李原原 译

上海远东出版社

丛书编委会

主　任：贾　峰

副主任：何家振　闫世东　郑立明

委　员：李原原　祝真旭　牛玲娟　梁雅丽　任泽林

　　　　王　岢　陈　卫　郑循如　吴建民　彭　勇

　　　　王梦雨　戴　虹　靳增江　孟　蝶　崔晓晓

特别感谢以下热心人士对童书工作的支持：

匡志强　方　芳　宋小华　解　东　厉　云　李　婧

刘　丹　熊彩虹　罗淑怡　旷　婉　杨　荣　刘学振

何圣霖　王必斗　潘林平　熊志强　廖清州　谭燕宁

王　征　白　纯　张林霞　寿颖慧　罗　佳　傅　俊

胡海朋　白永喆　韦小宏　李　杰　欧　亮

目录

Contents

一群驼鹿正穿过一片高山草甸，一只母驼鹿注意到有几只小驼鹿走得太靠近河边了。

　　"离河边远点儿，赶紧！"驼鹿妈妈朝小驼鹿们高声大叫。

A grazing herd of elk is moving across a mountain meadow when one of the cows notices a few calves moving too close to the river.

"Get away from that river bank, right now!" she shouts at them.

......一群驼鹿正穿过一片高山......

...herd of elk moves across a mountain ...

……很容易就会被灰狼抓住。

...where wolves can easily catch you.

"可是妈妈"，小驼鹿回答道，"这里的草，汁儿最多，还有边上这些柔软的柳叶……"

"如果你们冒险去那儿的话，很容易就会被灰狼抓住。"

"妈妈，你看这些河狸，他们多享受这片柳树林的树木和树皮，我们则可以享受那些柳叶。"

"But Mom," her calf replies, "this is where the juiciest grass grows. And just look at those soft willow leaves…"

"You are venturing into an area where wolves can easily catch you."

"But Mom, look at all those beavers, and how they are enjoying the wood and bark of this abundance of willows. And we could have all the leaves."

"河狸腿短，他们跑不过灰狼。但他们会游泳，还可以躲进自己的小屋然后逃走。现在请听我说，跟着我到安全的地方去。"

"这不公平！为什么河狸就可以住在这儿，一边伐树筑坝，给河水降温，帮助鱼类生存，一边享用美味的食物，而我们却必须到别处去吃草呢？"

"With their short legs, those beavers can't outrun a pack of wolves. But they can get away by swimming and hiding in their lodges. Please listen now, and follow me to where you will be safe."

"That's not fair! The beavers get to stay here, enjoying good food when cutting down trees to make dams that cool the stream so fish can survive, and we must go graze elsewhere?"

......他们可以通过游泳逃走......

... they can get away by swimming ...

但是如果我们不吃这些肥沃的草......

But if we do not eat these rich grasses ...

"我们不能吃那些河边的嫩草，太危险了。" 驼鹿妈妈回答说。

"但是如果我们不吃这些肥沃的草，它们就会死去。"

"这些草需要长在那儿，这样河岸才能保持坚固。快离开，免得灰狼抓住你们。"

"We should never graze on those soft grasses along the river. It is too dangerous," his mother replies.

"But if we do not eat these rich grasses, they will die off."

"Those long grasses need to grow there, for the river banks to remain strong. Get away from there now, before the wolves come and find you."

"哦，妈妈，您总是那么小题大做。别担心，我明白了。"

"仅仅明白不行，你们得赶紧离开！"

"好，好！我们过来了。谢谢，您总是教我们一些新东西。"

"Oh Mom, you are always so dramatic. Don't worry, I get the message."

"It is not enough that the message is understood – you have to act!"

"We will! We are coming. Thanks, you are always teaching us something new."

总是教我们一些新东西……

Always teaching us something new ...

……更多的树、更多的水、更多营养丰富的植物……

… trees, more water, more nutritious plants …

"看看周围，你们就会发现，自从灰狼回来以后，一切都变了。这里有了更多的树、更多的水，这意味着土壤中有了更多的矿物质，有了更多营养丰富的植物。这都要感谢灰狼。"

　　"让我们感谢灰狼？没门！"

"Look around you, and you'll discover all that has changed since the wolves came back. More trees, more water, and that means more minerals in the soil, and more nutritious plants. We have the wolves to thank for that."
"Thank the wolves? No way!"

"灰狼回来后，我们不敢在河岸上吃草，草地退化就放慢了，昆虫可以吃到更多的叶子。草地有了更多的老鼠和兔子，树上有了更多的鸟儿，土地就肥沃了。你们明白了吗？"

"明白了。明白了。但是，妈妈，这怎么能帮我每天都吃到美食呢？"

"自从灰狼回到这里，我们的数量确实增加了。你知道为什么吗？"

"Since they've returned, we've stopped grazing on the riverbanks, so there is less erosion, and more leafy food for insects. We have more mice and rabbits, and more birds in the trees, fertilising the ground. Do you get it?"

"I do. I do. But Mom, how does that help me get a good meal every day?"

"Our numbers have really increased since the wolves arrived. Do you know why?"

昆虫可以吃到更多的叶子……

More leafy food for insects ...

我们的森林现在有了更多的树

We now have more trees in our forests

"是因为您和爸爸聪明，每次他们想抓我们的时候我们都跑得比他们快吗？"

"哦，真希望我们跑得比灰狼快。实际上是我们的森林现在有了更多的树，因此就有了更多的食物。还有更多的河狸、兔子和老鼠。"

"听起来我们好像生活在乐园。" 小驼鹿说。

"难道不是吗？"

"Is it because you and Dad are clever, and outrun them every time they try to hunt us?"

"Oh, I wish that were true. We now have more trees in our forests, and so have more food. There are also more beavers, more rabbits, and more mice."

"It sounds like we live in a paradise," the calf says.

"Don't we?"

"您确定？周围有这么多灰狼，更别提灰熊和美洲狮了。"

"乐园是一个我们可以快乐健康地生活的地方，但这并不意味着不再有捕食者，或者让我们永远活着。乐园是一个我们都有机会做到最好的地方。"

……这仅仅是开始！……

"With so many wolves around? Not to even talk about all the grizzly bears and cougars?"

"Paradise is a place where we can live happy and healthy lives, but that doesn't mean we don't still have predators, or that we will live forever. It is a place where we all have a chance to be our best."

... AND IT HAS ONLY JUST BEGUN!...

AND IT HAS ONLY JUST BEGUN!

Did You Know ?

你知道吗?

When wolves are exterminated in a region, the elk population balloons five times, leading to overgrazing and the demolishing of woody vegetation, and destruction of birds' nests.

当一个地区的灰狼灭绝，驼鹿的数量将增加 5 倍，这会导致草场超载，木本植物被破坏，鸟巢被摧毁。

Scavengers thrive on the carcasses wolves leave behind, especially the more than 400 species of beetle that thrive on offal. Fungus further degrade carcasses and so increase nitrogen in the soil.

食腐动物以灰狼留下的尸体为食，尤其是 400 多种以内脏为食的甲虫。真菌会进一步降解尸体，从而增加土壤中的氮含量。

郊狼捕食叉角羚直至灭绝。但当有灰狼在附近时，郊狼就会远离叉角羚。叉角羚会在灰狼穴附近产仔，因为郊狼会避开这些区域。

Coyotes prey on pronghorn to the point of extinction. But when wolves are around, coyotes stay clear of the herds. Pronghorns will give birth to their fawns near wolf dens, since coyotes avoid these areas.

灰狼每天要游荡很远的距离，可达 20 千米。一只灰狼一顿可以吃 9 千克肉。灰狼群终其一生都聚在一起。

Wolves roam large distances, up to 20 km in a single day. One wolf can eat 9 kg of meat in a single meal. Wolves remain together in a pack all their lives.

Wolves are known to sacrifice themselves for the survival of the pack. The howling of the grey wolf is a penetrating sound. It is the primary communication tool between wolves.

众所周知，灰狼会为了群体的生存而牺牲自己。灰狼的嚎叫是一种具有穿透力的声音。它是灰狼之间主要的沟通方式。

While a cheetah can run at 100 km per hour over a short distance, a wolf is an endurance runner. This, combined with its intelligence, and its sense of hearing and smell, makes a wolf a superior hunter.

猎豹在短距离内的奔跑速度可达每小时 100 千米，而灰狼则是耐力型赛跑者。这一点，再加上它的智慧、听觉和嗅觉，使灰狼成为一名优秀的猎手。

The elk's antlers can grow up to one and a half centimetres per day. When fully grown these can weigh up to 40 kilograms. The growth of the antlers depends on sunshine, which boosts the elk's testosterone levels.

驼鹿的鹿角每天能长 1.5 厘米。完全成熟后，它们的重量可达 40 千克。鹿角的生长依赖于阳光，阳光可以提高驼鹿的睾丸素水平。

The anklebones of an elk make cracking or popping noises when they walk. Scientists suspect that elk use this sound to tell other elk that they're approaching them from behind.

驼鹿走路时，踝骨会发出砰砰的声音。科学家们怀疑驼鹿用这种声音告诉其他驼鹿它们正从后面靠近。

Think about It

想一想

Would you consider the arrival of a pack of wolves good news?

你认为灰狼的到来是好消息吗？

Should a mother leave her children to get on and discover things for themselves?

作为母亲，应该让孩子自己去探索和发现新事物吗？

What do beavers have to do with wolves?

河狸和狼有什么关系？

Is a place with grizzly bears, cougars and wolves a paradise?

一个有灰熊、美洲狮和灰狼的地方会是乐园吗？

Do It Yourself!

自己动手!

Our views on wolves may have been affected by children's stories that depict wolves in a negative light. It is time to change this perception, and to teach others abouthow they play a pivotal role in the ecosystem. What would you say are the most important features that make these superior hunters indispensable to the environment? List these features too. Ask others to also compile lists. Compare your lists, and discuss each item listed. Now write a report, documenting your findings.

我们对灰狼的看法可能受到儿童故事中对灰狼负面描写的影响。是时候改变这种观念了，让人们了解灰狼如何在生态系统中扮演关键的角色。你认为使这些优秀的猎人成为环境不可或缺的一分子的最重要的特征是什么？请列出这些特性，并请其他人和你一起编制清单。比较你们的清单，讨论清单中的每一项。现在写一份报告，记录你的发现。

学科知识
Academic Knowledge

生物学	驼鹿是最大的鹿科动物；驼鹿肉比牛肉或鸡肉含有更多的蛋白质；驼鹿的胃有4室，是反刍动物；柳树是一种落叶树，有排列成柔荑花序的小花，有非常强壮和发育良好的根，通常比茎大，使其抗风；柳树可以很容易地从折断的枝条上再生。
化 学	鲜草富含钾、膳食纤维、维生素A、维生素B1、维生素B2、维生素B3、维生素B5、维生素B6、维生素C、维生素E、维生素K、铁、锌、铜、锰、硒；柳树皮产生的水杨酸是阿司匹林的原料。
物 理	几千米外都能听到驼鹿的叫声；森林带来雨水，使当地气候凉爽；陆地上产生的70%的大气水分来自植物，这表明森林在跨大陆水循环中的重要性；树木释放出含有真菌孢子、花粉和微生物的气溶胶进入大气，这些微小颗粒为水的凝结提供凝结核，促进雨水形成；植物微生物组甚至帮助水分子在更高的温度下结冰，这是温带云层形成的关键一步。
工程学	河狸是"生态系统工程师"，因为它们建造的堤坝极大地改变了河流的特征（水流特征、地下水和形态）；河狸在自己的河流栖息地和生态系统中建造了一道又一道的堤坝。
经济学	创造更丰富的多样性有利于经济增长。
伦理学	关键物种的重新引入使生态系统恢复了活力。
历 史	驼鹿元素是许多美洲原住民部落的工艺品、音乐和故事中的特色。
地 理	白令海峡是更新世时期连接亚洲和北美的大陆桥；白令海峡为棕熊、骆驼、马、北美驯鹿、驼鹿以及人类提供了迁徙路线；森林补充了大气中的水蒸气。
数 学	系统动力学利用反馈循环和乘数效应跟踪区域的演化。
生活方式	驼鹿皮被用来制作圆锥形帐篷的覆盖物、毯子、衣服和鞋子的历史已经有数千年；父母在儿童教育中承担的角色。
社会学	在亚洲，鹿角被用作传统药材；拉科塔部落的男性会得到一颗驼鹿的牙齿作为长寿的祝福，因为牙齿是死驼鹿最后被分解的部分；柳树在中国象征着不朽与重生；在西方国家，柳树象征着悲伤。
心理学	父母传递给孩子的恐惧；学习如何评估风险（新鲜丰富的食物与死亡的风险）。
系统论	落基山脉的驼鹿亚种被重新引入，据估计所有北美亚种的数量超过了100万；河狸筑坝的技术可以为我们实现生态系统的重建和河流的回归提供新颖的、基于自然的解决方案；灰狼越多，掠食者就越多，但驼鹿也会变得更多。

情感智慧
Emotional Intelligence

母驼鹿

母驼鹿用她的威严来确保孩子的安全。她警告小驼鹿注意潜在的危险，并给出了精确的指令。她要求小驼鹿遵守她的规定，并指明了安全地带。她不准备花长时间去争论，也不准备详细解释。她耐心告诫小驼鹿，安全总比后悔好。虽然她想让小驼鹿学会谨慎，但她也想让小驼鹿去积极思考为何灰狼回来后当地的生态系统会恢复生机。这让她有机会分享关于乐园的智慧，指出乐园是所有物种都可以过着幸福健康生活的地方。

小驼鹿

小驼鹿不准备听从妈妈的命令。他看到丰富的食物，不明白为什么不能拥有这些食物。他觉得妈妈的命令不公平。他承认妈妈很有智慧，但他仍然认为，妈妈对灰狼起到的积极作用表达感激，这实在有些夸张。虽然他明白了狼群的积极作用，但他不理解为什么要感谢灰狼。他没有抓住父母逃脱狼群猎杀的关键，也没有领会这样一个事实：灰狼的存在有助于提高生活质量。他的观点代表了大多数人的观点，他们不了解灰狼为该地区所有物种所作的生态贡献。

艺术
The Arts

灰狼是绘画中很受欢迎的主题，除了外表，它们还有更多值得关注的地方。灰狼的嚎叫和外表一样迷人。你能通过嚎叫表达情感吗？在互联网上找到狼叫的录音，然后模仿它们的嚎叫。当你掌握了基本技巧，就开始用不同方式的嚎叫来传达不同的信息吧。录下你的嚎叫，让你的朋友和家人听听录音，不要告诉他们这是你的声音。他们欣赏你"与狼对话"的艺术吗？

思维拓展
Systems: Making the Connections

在生命之网中，一切都是相互联系的，一旦一个基础物种从生态系统中消失，整个系统都会受到影响。灰狼在陆地上的作用和鲸在海洋中的作用类似。这些物种不仅影响生态系统的化学和生物学过程，如土壤中的氮或水中的铁，而且还影响生态系统中物种的数量。正是通过这些关键物种，多样性才得以维持，甚至恢复。灰狼留下的尸体为400多种昆虫提供了营养。而昆虫是受滥用杀虫剂和除草剂影响最大的物种。值得注意的是，当灰狼不再被子弹射杀，鲸不再被鱼叉刺穿时，生态金字塔底部的物种才会兴旺。我们几乎很少注意陆地上丰富多样的昆虫和海洋里的浮游生物。我们可以对不同的生态系统进行比较。我们需要多样性，还需要知道如何确保生态系统的营养级联。即使我们知道的还不够多，也不能完全理解每个物种在生命之网中的作用，我们也必须清楚认识这样一个事实，即每个物种都在尽自己最大的能力作贡献。正是通过他们的共同努力，形成了一个有韧性的生态系统。虽然我们倾向于尊重大的和可见的，但是生态系统的整体力量也依赖于最小的和不可见的物种。这使我们认识到，"乐园"的创造不仅仅是体验快乐，那里的生活不是由猎物和捕食者之间的仇恨决定的。乐园的定义，是指在一个生态环境中，每个物种都有机会发挥自己最好的一面，过上最好的生活。狼群的存在使得驼鹿分散成更小的群体，分布得更广，从而更好地利用生态空间。在这里没有必要寻求规模经济。相反，每个物种都在一群互补物种中不断调整自己的生存范围，使得所有物种都能受益。

动手能力
Capacity to Implement

对你生活的地区的生物多样性状况做一些研究。想想有哪些植物、动物、微生物和你生活在同一地区？哪些物种正在消失，它们是如何消失的？咨询专家，并根据你生活的地区的具体情况，想想需要重新引入哪些大型动物以便使生态系统恢复生机。一旦你确定了那些失踪的"生命的主人"，制定一个计划并寻求他人的帮助来重新引入这些物种。

故事灵感来自

This Fable Is Inspired by

多莉·乔金森
Dolly Jørgensen

1994 年，多莉·乔金森获得得克萨斯农工大学土木工程和环境工程专业的学士学位。2003 年，她通过研究中世纪的林业和渔业管理获得休斯敦大学的硕士学位。2008 年，她通过研究英国和斯堪的纳维亚城市卫生获得弗吉尼亚大学的博士学位。乔金森博士也研究人们帮助动物回归、复苏，甚至复活已灭绝物种等行为背后的历史和情感因素。她也是《恢复现代失去的物种》（麻省理工学院出版社）一书的作者。乔金森博士目前是挪威斯塔凡格大学的历史学教授。

图书在版编目（CIP）数据

冈特生态童书.第七辑:全36册:汉英对照 /
（比）冈特·鲍利著;（哥伦）凯瑟琳娜·巴赫绘;
何家振等译.—上海:上海远东出版社,2020
ISBN 978-7-5476-1671-0

Ⅰ.①冈… Ⅱ.①冈… ②凯… ③何… Ⅲ.①生态
环境–环境保护–儿童读物—汉英 Ⅳ.①X171.1-49

中国版本图书馆CIP数据核字（2020）第236911号

策　　划　张　蓉
责任编辑　祁东城
封面设计　魏　米　李　廉

冈特生态童书
河流新生机
[比]冈特·鲍利　著
[哥伦]凯瑟琳娜·巴赫　绘
李原原　译

记得要和身边的小朋友分享环保知识哦！
八喜冰淇淋祝你成为环保小使者！